KU-769-116

# BIOLOGY PRACTICAL GUIDE 2

## CHEMICAL REACTIONS IN ORGANISMS

Revised Nuffield Advanced Science
Published for the Nuffield–Chelsea Curriculum Trust
by Longman Group Limited

**Longman Group Limited**
Longman House, Burnt Mill, Harlow, Essex CM20 2JE, England
and Associated Companies throughout the World

First published 1970
Revised edition first published 1985
Copyright © 1970, 1985. The Nuffield–Chelsea Curriculum Trust

Design and art direction by Ivan Dodd
Illustrations by Oxford Illustrators

Set in Times Roman and Univers
and made and printed in Great Britain
by The Bath Press, Avon

ISBN 0 582 35428 5

All rights reserved. No part of this publication may be reproduced,
stored in a retrieval system, or transmitted in any form or by any
means – electronic, mechanical, photocopying, or otherwise –
without the prior written permission of the Publishers.

**Cover photograph**
Transverse section of the ileum of a cat ($\times 110$). See investigation 6D
'The fine structure of the intestinal wall'.
*From Freeman, W. H. and Bracegirdle, B.*, An atlas of histology, *2nd
edn, Heinemann, 1967.*

# CONTENTS

# SAFETY

In these *Practical guides*, we have used the internationally accepted si; given below to show when you should pay special attention to safety.

 highly flammable

 explosive

 toxic

 corrosive

 radioactive

 take care! (general warning)

 risk of electric shock

 naked flames prohibited

 wear eye protection

 wear hand protection

# INTRODUCTION

The practical investigations in this *Guide* relate largely to the topics covered in *Study guide I*, Part One, 'Maintenance of the organism', Chapters 5 to 7. Cross references to the *Study guide* are given.

**CELLS AND CHEMICAL REACTIONS**

**Investigation 5A A metabolic pathway in yeast.** (*Study guide* 5.6 'Metabolism, the Laws of Thermodynamics, and the role of enzymes'.)
The rate of fermentation of yeast is examined, using different substrates and at different temperatures.

**Investigation 5B Enzyme extraction.** (*Study guide* 5.6 'Metabolism, the Laws of Thermodynamics, and the role of enzymes'.)
An enzyme extract from germinating barley grains is prepared and tested.

**Investigation 5C The course of an enzyme-catalysed reaction.** (*Study guide* 5.6 'Metabolism, the Laws of Thermodynamics, and the role of enzymes'.)
A colorimeter is used to follow the course of the digestion of starch.

**Investigation 5D An enzyme-catalysed synthesis.** (*Study guide* 5.6 'Metabolism, the Laws of Thermodynamics, and the role of enzymes'.)
An enzyme extract is prepared from potatoes and is then tested for its ability to catalyse the synthesis of starch.

**Investigation 5E The uptake of oxygen as a measure of metabolism.** (*Study guide* 5.7 'Cellular respiration and the mitochondrion'.)
The respiratory metabolism of germinating mung beans is investigated, using a respirometer.

**Investigation 5F Respiratory quotient.** (*Study guide* 5.7 'Cellular respiration and the mitochondrion'.)
A respirometer is used to provide data from which the respiratory quotient is calculated.

**HETEROTROPHIC NUTRITION**

**Investigation 6A Digestion by micro-organisms and tissues.** (*Study guide* 6.4 'Digestion'.)

Starch agar plates are used to indicate when digestion is taking place.

**Investigation 6B Digestive organs: a model gut.** (*Study guide* 6.5 'The double function of the alimentary canal'.)
Visking tubing is used to demonstrate the function of the gut.

**Investigation 6C Digestion and absorption in the gut of a mammal.**
(*Study guide* 6.5 'The double function of the alimentary canal'.)
The alimentary canal is examined macroscopically and microscopically. Chromatography is used to investigate the products of digestion.

**Investigation 6D The fine structure of the intestinal wall.** (*Study guide* 6.7 'Absorption'.)
A microscopic examination relating structure to function.

**Investigation 6E A microscopic investigation of the liver.** (*Study guide* 6.8 'The role of the liver'.)

## Chapter 7  PHOTOSYNTHESIS

**Investigation 7A The interaction of plants and animals.** (*Study guide* 7.1 'Photosynthesis and the biosphere'.)
Artificial terrestrial and aquatic communities are set up in sealed, transparent containers to study the way in which organisms affect the atmosphere surrounding them.

**Investigation 7B The structure of a leaf.** (*Study guide* 7.2 'The sites of photosynthesis'.)
A microscopic examination of leaves and the way in which the structural features relate to their functions.

**Investigation 7C The evolution of oxygen.** (*Study guide* 7.3 'The mechanism of photosynthesis'.)
The rate of production of oxygen by a water plant at various light intensities is investigated.

**Investigation 7D Leaf pigments.** (*Study guide* 7.3 'The mechanism of photosynthesis'.)
Leaf pigment extracts are prepared and analysed, using either thin layer or paper chromatography.

**Investigation 7E The reducing activity of chloroplasts: 'the Hill reaction'.** (*Study guide* 7.3 'The mechanism of photosynthesis'.)
The reducing power of isolated chloroplasts is investigated.

**Investigation 7F  The production of starch by leaves.** (*Study guide* 7.5 'The reduction of carbon dioxide'.)
Some of the conditions that are required for the synthesis of starch in a plant's leaf are investigated.

**Investigation 7G  Carbon fixation in CAM plants.** (*Study guide* 7.6 'Crassulacean acid metabolism – the CAM pathway'.)
Leaf extracts of light-treated and dark-treated CAM and non-CAM plants are prepared and compared.

## A note for users of this *Practical guide*

The instructions given for the investigations are intended for use as guidelines only. We hope that you will modify and extend the techniques that have been described to meet your own requirements. Other organisms should certainly be tried, depending on what is most readily available. Some of these investigations may lend themselves to further work in a Project.

It may not always be possible, for various reasons, for you to do a practical investigation for yourself. A study of data from another source is perfectly acceptable in such a case.

# CHAPTER 5

# CELLS AND CHEMICAL REACTIONS

### INVESTIGATION
### 5A  A metabolic pathway in yeast

(*Study guide* 5.6 'Metabolism, the Laws of Thermodynamics, and the rol
of enzymes'.)

Many yeasts of the genus *Saccharomyces* are facultative anaerobes. Thi
means that, in the presence of oxygen, they will oxidize a carbon source
such as glucose. Carbon dioxide and water are produced. On the othe
hand they can grow in the absence of oxygen. When this happens a forn
of respiration occurs in which the sugar is converted to ethanol and
carbon dioxide. This conversion is usually called a *fermentation*. It i
brought about by a sequence of more than fourteen chemical reactions i
a *metabolic pathway*.

The rate at which carbon dioxide is produced can be used as a
measure of the overall rate of fermentation. This investigation examine
how the rate of fermentation varies with temperature and with the type o
sugar being metabolized.

The procedure describes two separate experiments which can b
done side by side. If time is limited these experiments should be carrie
out by different groups. The first experiment (steps **1–8** below) examine
how the rate of fermentation varies with the type of sugar bein,
metabolized. The second experiment (steps **9–14** below) examines ho
fermentation varies with changes in temperature.

*Procedure*
Note: Steps **1–3** and **9** must be set up at least an hour before the rest o
the work can be done.

*The fermentation of different substrates*
1   Prepare a yeast suspension by stirring 4 g of dried yeast and 1 g of
    yeast extract into $100 \, cm^3$ of water.
2   Start a yeast culture by thoroughly mixing $20 \, cm^3$ of the yeast
    suspension with $20 \, cm^3$ of glucose solution ($0.2 \, mol \, dm^{-3}$). (This
    volume should be sufficient for twelve experiments.) Incubate this in
    a water bath at 35 °C for at least one hour to let fermentation start.
3   Repeat step **2** with any other available fermentation substrates.
    These might include sugars such as galactose, which is very similar
    to glucose, and fructose, lactose, and sucrose. *Include a suitable
    control.*
4   Place two clean test-tubes next to each yeast culture tube in the

water bath. Transfer 2–3 cm³ of the culture into each test-tube.

5  Next, a small Durham tube has to be filled with yeast culture and inverted in each test-tube. To do this follow the instructions in *figure 1*.

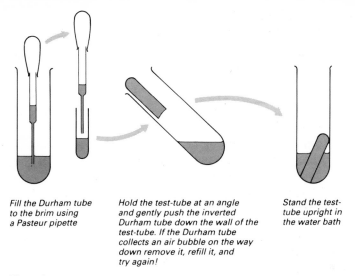

Fill the Durham tube to the brim using a Pasteur pipette

Hold the test-tube at an angle and gently push the inverted Durham tube down the wall of the test-tube. If the Durham tube collects an air bubble on the way down remove it, refill it, and try again!

Stand the test-tube upright in the water bath

**Figure 1**
Filling and inverting a Durham tube.

6  Note the time at which each Durham tube is set up. At timed intervals (every five minutes should be suitable), measure the length of the bubble collecting inside each tube.

7  Pool the results obtained by the whole class. Work out the mean bubble length at each time interval. Plot a graph of the mean bubble length against time for each sugar.

8  By reference to textbooks and molecular models, work out the *differences* between the structures of the sugar molecules you have used in this investigation. (Avoid recording *all* the details of these molecular structures.)

*The temperature dependence of fermentation*

9  Set up three water baths at temperatures between 20 °C and 50 °C (*e.g.* 20 °C, 35 °C, and 50 °C). Set up a culture of yeast at 35 °C containing: 2 g dried yeast, 3 g glucose, 1 g yeast extract, 100 cm³ water. Incubate for at least one hour to allow fermentation to start.

10  Transfer 2–3 cm³ of the culture to each of six test-tubes. Place two of these in each water bath. Allow them to equilibrate for five minutes.

11  Follow the instructions in *figure 1* to fill Durham tubes and insert them upside down into each test-tube. (Do not remove the test-tubes from their water baths for too long.)

**12** Record the length of the carbon dioxide bubbles inside the Durham tubes at timed intervals.

**13** Pool the results obtained by the whole class. Work out the mean bubble length at each time interval. Plot a graph of mean bubble length against time for each temperature.

**14** If your data, at a given temperature, lie reasonably close to a straight line estimate the *rate* of fermentation by measuring the slope of this line. Calculate the $Q_{10}$ for fermentation as follows:

$$Q_{10} = \frac{\text{rate of fermentation at } (t + 10)\,^{\circ}\text{C}}{\text{rate of fermentation at } t\,^{\circ}\text{C}}$$

*Questions*

**a** *Do your results suggest that the metabolism of yeast cells shows a preference for one sugar rather than another? If so, which sugar is preferred, and what features of its molecular structure distinguish it from the other sugars which you have tested?*

**b** *If a yeast cell were presented with several sugars in its environment, what type of mechanism might ensure that it would absorb most rapidly the sugar that it could metabolize most efficiently?*

**c** *In what way does the rate of fermentation vary with temperature? Though the data obtained in this investigation are semi-quantitative it may be possible to estimate the $Q_{10}$ value for fermentation. Does your estimate agree with what you expect for physiological processes?*

**d** *Summarize the evidence that you have obtained to suggest that an enzyme (or enzymes) could be involved in fermentation.*

**e** *Cell-free extracts of yeast can carry out alcoholic fermentation. That is, they will catalyse the conversion of sugars into ethanol and carbon dioxide in the absence of living cells. Suggest how you might find out whether such an extract contains a single enzyme or several enzymes responsible for fermentation.*

**INVESTIGATION**
**5B Enzyme extraction**

(*Study guide* 5.6 'Metabolism, the Laws of Thermodynamics, and the role of enzymes'.)

Metabolic processes, such as fermentation, involve several chemical reactions, each catalysed by a specific enzyme. If a particular reaction in the process is to be studied in detail two things must be done. Firstly, the relevant enzyme must be extracted from the tissues in which it is found. Secondly, a method of measuring the course of the reaction must be devised. Since it is relatively easy to measure the concentration of starch suspensions (see investigation 5C), this and the next two investigations examine starch-metabolizing enzymes. Seeds commonly have starch as a food store and we might expect to find enzymes which will break down starch when the carbohydrate is needed during germination. In this investigation an extract from germinating barley grains is prepared and then tested for its ability to break down starch.

*Procedure*
Use 8 g of germinating barley grains to make 100 cm$^3$ of extract. For each experiment you will require 10 cm$^3$ of extract.

1  Immerse the barley grains in a 1% solution of sodium chlorate(I) (sodium hypochlorite) for three minutes to surface sterilize. Pour away the solution and wash the grains twice in sterile water.

2  Cover the grains with sterile water and leave for twenty-four hours at 25 °C.

3  Pour off the water and arrange the grains in a thin layer on wet filter paper. Cover to prevent drying out and leave at 25 °C for a further twenty-four hours or until the majority of grains have roots protruding.

4  Crush or grind the grains in distilled water. (At this stage it is advisable to use less than the required 100 cm$^3$ of water per 8 g of grains. The remaining water can then be used to rinse the crushed grains out of the apparatus.) Use either a pestle and mortar or a mechanical homogenizer (blender). With the latter, see that the blades are covered with water. Continue grinding or homogenizing just long enough to ensure that no intact grains remain. (See *figure 2*.)

5  Make sure that the total volume of the extract is 100 cm$^3$ for every 8 g of barley grains used. If necessary make up to this volume with more water.

6  Test this crude extract to see if it digests starch. Take a small sample and add it to 1 cm$^3$ of starch suspension (0.1%). After ten minutes test by adding one drop of iodine solution (iodine dissolved in an aqueous solution of potassium iodide).

7  Filter the extract to remove the solid debris. Drape three or four layers of muslin or nylon stocking across the mouth of the beaker. Slowly pour the extract into the centre. Gather up the corners of cloth and twist so that most of the liquid can be squeezed out. Test this filtrate for ability to digest starch (see step **6**).

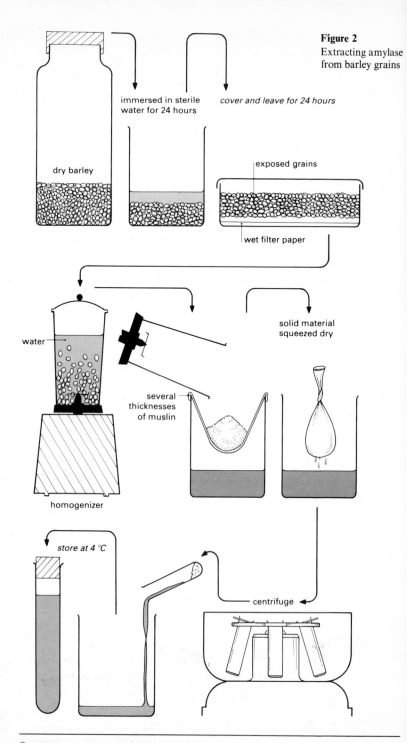

**Figure 2**
Extracting amylase from barley grains

immersed in sterile water for 24 hours

*cover and leave for 24 hours*

dry barley

exposed grains

wet filter paper

water

solid material squeezed dry

several thicknesses of muslin

homogenizer

*store at 4 °C*

centrifuge

**8** Pour the milky fluid into centrifuge tubes, being careful to put equal volumes into each tube. Check that the tubes are properly balanced in the bench centrifuge. Spin at maximum speed for five minutes. Pour the clear liquid (supernatant) into a suitable container (a stoppered conical flask or boiling-tube) and discard the solid material.

**9** Test a sample of the extract as before (see step **6**) and store the remainder at 4 °C (in a refrigerator).

*Questions*

**a** *Which parts of the germinating barley, do you think, have been discarded in the extraction process?*

**b** *What tests could be performed on the discarded material to reveal why the seeds should contain starch-digesting enzymes?*

**c** *How would you obtain evidence that the starch-digesting activity of the extract is indeed due to the presence of an enzyme (that is, an amylase)?*

**d** *Many enzymes are likely to denature (lose their specific protein structure and enzyme activity) when they are removed from the special conditions of the cytoplasm. Were any special precautions taken to ensure that the barley extract retained its ability to degrade starch?*

## INVESTIGATION
**5C The course of an enzyme-catalysed reaction**

(*Study guide* 5.6 'Metabolism, the Laws of Thermodynamics, and the role of enzymes'.)

We know that the enzyme extract prepared in the previous investigation is capable of digesting starch, but we have little detailed knowledge of the reaction and the enzyme which catalyses it. We cannot progress further without a quantitative method of measuring the course of the reaction. Fortunately we can take advantage of the fact that starch forms a coloured compound with iodine solution. The intensity of this colour will be proportional to the concentration of the compound. Therefore, a method of measuring the intensity of its colour will provide a measure of the concentration of starch in the suspension.

The instrument commonly used to measure colour intensity is called a *colorimeter*. It has a light source, a chamber to hold a test-tube of the liquid being studied, and a photo cell connected to a meter (see *figure 3* ).

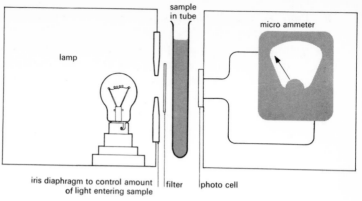

**Figure 3**
A colorimeter.

As the light passes through the liquid the coloured compound will absorb some of it. The denser the colour, the less the amount of light transmitted through the liquid and therefore the smaller the reading on the meter scale. In this way the concentration of a suspension of starch can be estimated from the meter reading. However, before this can be done it will be necessary to calibrate the meter readings against a series of starch suspensions of known concentration (see steps **10–12** in the procedure that follows).

An important part of any colorimeter is the light filter and its function should be understood. Blue light passes through a blue liquid; red and yellow light are absorbed by it. Thus blue light is the least useful when trying to measure the degree of blueness of a liquid; red light is the most useful. For this reason a red filter is used in the colorimeter when measuring starch suspensions. Before starting the following procedure make yourself thoroughly familiar with the workings of the colorimeter and make sure that you can get consistent readings for an unchanging sample of starch/iodine.

*Procedure*

1   Mix standard iodine solution with distilled water in the ratio 1:20. (For example, 0.5 cm³ iodine:10 cm³ water.) The final volume will depend on the size of the sample tube for your colorimeter. This tube should be more than half full with the diluted iodine solution. Swirl to mix well.

2   Insert the tube into the colorimeter and switch on the meter. Rotate the sample tube to give the greatest possible deflection of the meter needle. Put a mark on the side of the tube to indicate this position in the colorimeter. For subsequent readings always use the sample

tube, placed in the colorimeter in this same position. (Wash the tube well in between readings.)

3   With the sample tube still in the colorimeter adjust the meter reading to give full-scale deflection. At this point the meter should read either 100 per cent (transmission) or zero optical density, depending on how the scale is labelled.

4   Place exactly $15\,cm^3$ of a 0.2 per cent suspension of starch into a tube or vessel and put this into a water bath at $25\,°C$. (Later you will have to withdraw small samples from this vessel. Make sure that your syringe or pipette will reach to the bottom.)

5   Place exactly $5\,cm^3$ of the enzyme extract into a test-tube and incubate this in the same water bath.

6   When the starch suspension and enzyme extract have equilibrated with the water bath (after about ten minutes) mix them quickly together and swirl thoroughly. Start a stopclock and return the reaction mix to the water bath. (See *figure 4*.)

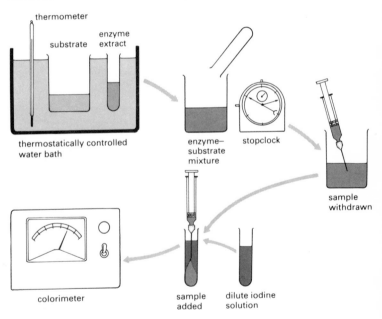

**Figure 4**
The procedure for following the course of an enzyme reaction.

7   As quickly as possible remove a carefully measured sample of the reaction mix and add it to the dilute iodine solution in the sample tube. Decide on the volume of your sample as follows:

*Volumes to be used in the sample tube:*

| | Iodine solution | Water | Reaction mix |
|---|---|---|---|
| Ratio | 1 | 20 | 1 |
| Example | $1 \text{ cm}^3$ | $20 \text{ cm}^3$ | $1 \text{ cm}^3$ |
| Volumes | $0.5 \text{ cm}^3$ | $10 \text{ cm}^3$ | $0.5 \text{ cm}^3$ |

Swirl to mix. Record the time on the stopclock and the reading on the colorimeter.

**8** As soon as this reading has been taken discard the contents of the sample tube, wash it, and put in fresh iodine solution as in step **1**. Remove a second sample from the reaction mix and obtain a colorimeter reading as in step **7**.

**9** Continue taking samples and recording their colorimeter readings as quickly as possible. If the colour intensity changes by a large amount over the first three or four readings carry on taking samples and readings as fast as you can. If there is little change in the colorimeter reading from one sample to another then take longer time intervals in between samples (for instance every five minutes).

**10** Convert the meter readings into units of starch concentration as follows. Put fresh iodine solution into the sample tube as in step **1**. Add the typical sample volume of starch suspension (0.2 per cent). Swirl well and obtain a colorimeter reading for this mixture. Repeat for each of the following dilutions of the starch suspension:

| Starch concentration (%)* | 0.14 | 0.1 | 0.07 | 0.04 |
|---|---|---|---|---|
| Starch suspension (0.2 %), cm$^3$ | 7 | 5 | 3 | 2 |
| Distilled water, cm$^3$ | 3 | 5 | 7 | 8 |

* concentration of $1\% = 1 \text{ g per } 100 \text{ cm}^3$

**11** Plot a graph of the readings (vertical axis), against the starch concentration (horizontal axis).

**12** Use this graph to convert the first series of readings into their equivalent starch concentrations. (See *figure 5*.)

*Questions*

**a** *Draw a graph of starch concentration against time and compare it with figure 6. In what ways are the two graphs similar? How do they differ?*

**b** *Look up the logarithm (base 10) of each starch concentration and then plot the log of starch concentration against time. (Before doing this it is a good idea to express the starch concentrations in mg $100 \text{ cm}^{-3}$. This avoids decimals and the corresponding negative logarithms.) What information does this graph provide which the previous one did not?*

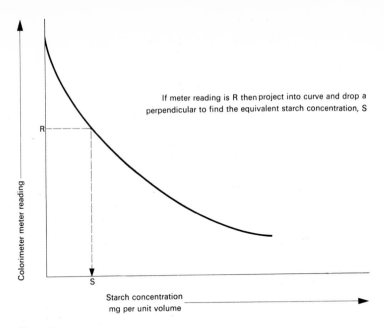

If meter reading is R then project into curve and drop a perpendicular to find the equivalent starch concentration, S

**Figure 5**
The use of a conversion graph.

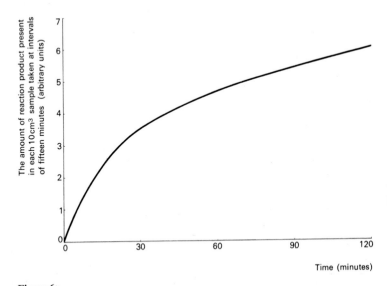

**Figure 6**
The progress of the digestion of a protein (casein).
*After Baldwin, E.,* Dynamic aspects of biochemistry, *Cambridge University Press, 1947, and Baldwin, E.,* The nature of biochemistry, *Cambridge University Press, 1967.*

c   *Name one phenomenon, not necessarily biological, which shows a similar relationship with time. What is the reason for the similarity between starch digestion and the phenomenon that you have named?*

d   *How would you modify the above procedure to investigate either the effect of temperature (0 ° to 100 °C) or the effect of the pH of the reaction mixture on starch digestion? Give sufficient details to serve as instructions.*

e   *Suggest a hypothesis to explain what happens to the starch when the suspension is mixed with the enzyme extract. What experiments would test this hypothesis?*

### INVESTIGATION
**5D   An enzyme-catalysed synthesis**

(*Study guide* 5.6 'Metabolism, the Laws of Thermodynamics, and the role of enzymes'.)

Inside cells there is a balance between the reactions that break substances down and the reactions that build up the components of cells. All of these intracellular reactions are catalysed and controlled by enzymes. No practical investigation of enzyme action would be complete without reference to at least one enzyme-catalysed synthesis.

After dealing with the digestion of starch it is appropriate to ask how starch is formed in the first place. Not surprisingly the tissue of potato tubers turns out to be a good source of the enzyme(s) responsible for the synthesis of starch. Since the digestion of starch yields first maltose and then glucose it could be that the synthesis of starch is a simple reversal of this process. Alternatively, there are a number of compounds inside cells which are closely related to the simple sugars. One example is glucose 1 phosphate (see *figure 7*) and it could be that this provides the building blocks of starch molecules.

This investigation makes use of the techniques developed in earlier investigations (5B and 5C) for enzyme extraction and measuring starch concentration. The aim is to prepare an extract of potato tuber and check for its ability to catalyse the synthesis of starch. In doing this it will be necessary to discover what compounds are suitable as precursors, or substrates, for starch synthesis.

An extract of potato tubers is bound to contain a great deal of starch. All of this starch must be removed before we can test for the extract's ability to catalyse the synthesis of new starch. Fortunately starch is stored inside cells in relatively large granules which will sediment during centrifugation.

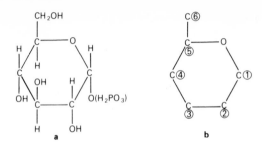

**Figure 7**
Glucose 1-phosphate.
**a** Structural formula;
**b** The method of numbering the carbon atoms.

*Procedure*
Note: The enzyme extraction, steps **1–5**, will take about half an hour. The extract can be stored overnight at 4 °C, though it will turn brown. If the whole investigation is to be completed in one session the substrate solutions and the colorimeter must be prepared for use (steps **6, 7,** and **11**) while the enzyme extract is being made.

1 To prepare the enzyme extract for one experiment take two medium-sized potatoes, peel and cut into small pieces. Crush these with a pestle and mortar or use a mechanical homogenizer. Add water sparingly so that the resulting mash is just liquid enough to be poured from its container.

2 Pour the crushed potato quickly through a single layer of muslin or stocking.

3 Transfer the extract to centrifuge tubes. Spin it in a bench centrifuge for five minutes at the highest speed to separate the starch granules.

4 When the centrifuge head has stopped spinning take one drop of clear liquid from the top of each tube. Test each drop on a white tile with iodine solution. If the blue colour characteristic of starch appears, centrifuge for a further period of five minutes.

5 Repeat the iodine test and, if necessary, continue to centrifuge until no starch is detectable in the samples of clear liquid. Once this has been achieved, carefully pour the clear liquid from each centrifuge tube into a single container. If this is to be kept for some time before using, stopper the container and place it in a refrigerator.

6 Prepare a solution of 0.5 g of glucose 1-phosphate dissolved in 50 cm$^3$ of distilled water (a 1 per cent solution). This will be sufficient for several experiments. The compound is unstable in solution, hydrolysing fairly rapidly to glucose and phosphoric acid at room temperature. It should be prepared just before use or stored in a refrigerator.

**7**    Prepare other substrate solutions of the same concentration, that is 1 per cent, of glucose, maltose, and sucrose.

**8**    Put 5 cm³ of the glucose 1-phosphate solution into a test-tube and place it in a water bath set at 25 °C. Do the same with the other substrate solutions.

**9**    Take three 5 cm³ samples of the clear potato extract and put each in a test-tube in the same water bath.

**10**    Pour either 10 or 20 cm³ of distilled water into a colorimeter tube. (The volume chosen must fill more than half of the tube.) Add 0.5 or 1 cm³ of iodine solution and swirl to mix well.

**11**    Set up the colorimeter using a red filter as before (see investigation 5C).

**12**    Add the 5 cm³ of potato extract to each tube containing a substrate solution. Mix each very thoroughly and start the stopclock.

**13**    Take a sample from one of the mixtures and add it to the iodine solution in the colorimeter tube. The volume of the sample should be 0.5 or 1 cm³, depending on the size of the colorimeter tube. With this in the colorimeter set the meter reading to full-scale deflection (100 per cent transmission or zero optical density). This provides a zero reading, that is a reading with no starch present, as a reference for future measurements. Discard and prepare a fresh iodine solution in the colorimeter tube.

**14**    After two minutes take a single drop of the extract/glucose 1-phosphate mixture and mix it with a drop of iodine solution on a white tile. Look for a very slight colour change. If there is none repeat the operation at two minute intervals as recorded by the stopclock.

**15**    As soon as the slightest trace of blue or grey colour appears on the tile, take a precise sample volume (i.e. either 0.5 or 1 cm³) from the extract/glucose 1-phosphate mixture. Add this to the iodine solution in the colorimeter tube. Mix well and record the meter reading and the time.

**16**    Using fresh iodine solution carry out similar measurements with the other three extract/substrate mixtures.

**17**    Check the amount of enzyme/substrate mixture left and arrange to take more samples at regular intervals. Try to take at least eight samples for each extract/substrate mix. If the meter readings are changing rapidly from one sample to the next with a particular substrate, concentrate on that reaction mix and take samples as often as possible. Then return to the other extract/substrate mixtures.

**18**    For each measurement record the name of the substrate, time, colour, and the meter reading.

**19** Convert the meter readings into starch concentrations using the conversion graph described in investigation 5C *figure 5*.

**20** Construct graphs for the progress of each reaction, plotting starch concentration against time.

*Questions*

**a** *Describe the appearance of the mixtures at the end of the investigation. Which of the substrates produced starch after the addition of potato extract?*

**b** *Compare the graphs obtained in this investigation with that of starch digestion which you drew in answer to question a of investigation 5C. Describe any similarities and differences between them.*

**c** *Can you detect from the graphs any differences in the early phases of starch synthesis and digestion? If so, suggest a hypothesis to account for this difference.*

**d** *What evidence have you obtained which will enable you to deduce whether or not starch synthesis is a simple reversal of starch digestion?*

**e** *Make a list of likely substrates for the potato enzyme which might be tested in addition to glucose, maltose, and sucrose.*

**INVESTIGATION**

**5E The uptake of oxygen as a measure of metabolism**

(*Study guide* 5.7 'Cellular respiration and the mitochondrion'.)

The previous three investigations have been concerned with the activity of certain enzymes extracted from their cells. The next two investigations return to the metabolism of whole organisms. As seen in investigation 5A, measurements of gaseous exchange can give a clear indication of the activity of respiratory metabolism. This investigation describes a quantitative method of making such measurements. Techniques of this type have been immensely important in unravelling the chemical details of respiratory metabolism.

The respirometer is used here to measure the uptake of oxygen by respiring mung beans. Any other small living organisms could be used for this investigation. This is done simply by measuring the change in the volume of gas surrounding the beans. It is, therefore, essential to eliminate volume changes that are caused by anything other than uptake of oxygen. As the beans respire they will, of course, produce carbon dioxide. This is absorbed in a solution of potassium hydroxide.

The apparatus (see *figure 8*) consists of two vessels, one containing the organisms and the other acting as a thermobarometer. Small changes in temperature or pressure cause air in this second vessel to expand and contract. This opposes and compensates for similar changes in the first vessel. Changes in the manometer level are thus due only to the activities of the organisms themselves.

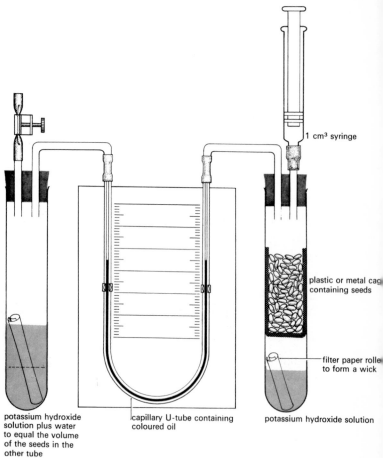

1 cm³ syringe

plastic or metal cage containing seeds

filter paper rolled to form a wick

potassium hydroxide solution

potassium hydroxide solution plus water to equal the volume of the seeds in the other tube

capillary U-tube containing coloured oil

**Figure 8**
A respirometer set up to measure the uptake of oxygen by seeds.

*Procedure*

1  Pour 5 cm³ of potassium hydroxide solution (15%) into both respirometer vessels; use a funnel so that none touches the sides of the vessels.
2  Add small rolls of filter paper to act as wicks.
3  Fill the basket or cage with beans and put it into a vessel. Make

sure that the beans are not touching the potassium hydroxide solution or wick.

4  Fit one bung with a 1 cm³ syringe and connecting tube, as shown in *figure 8*. Fit a second bung with a screw clip and connecting tube.

5  Estimate the total volume of beans and basket together and add this amount of water to the other respirometer vessel.

6  Draw coloured kerosene or Brodie's fluid into the manometer U-tube. The fluid must be free from bubbles and come to about the middle of the scale on each side.

7  Open the screw clip and remove the syringe; then connect the manometer U-tube.

8  Place the manometer so that both respirometer vessels are immersed in a water bath maintained at 20 °C. The manometer should be suspended outside (see *figure 9*).

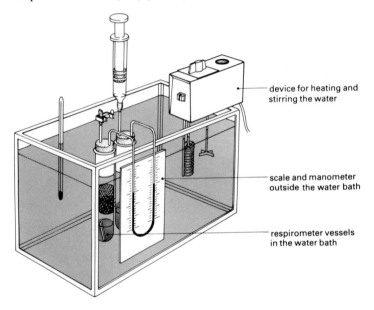

device for heating and stirring the water

scale and manometer outside the water bath

respirometer vessels in the water bath

**Figure 9**
The respirometer in use.

9  Set the piston of the syringe at about the 0.5 cm³ mark and, when the respirometer has been in the water bath for about five minutes, insert the syringe as shown. Close the screw clip. By means of the syringe adjust the manometer so that the fluid levels are the same on both sides.

10  Record the exact positions of the syringe piston, of the menisci in both sides of the manometer, and the time.

**11** Record new positions of the manometer fluid at four-minute intervals. When it nears the end of the scale on one side restore it to its original position and note the new position of the syringe piston.

**12** Plot a graph of meniscus level against time. Continue to take readings until four consecutive ones lie on the same straight line.

**13** Raise the temperature of the water bath to 30 °C and repeat steps 9–12.

**14** Remove and weigh the beans.

*Questions*

**a** *When the graph of manometer readings against time is a straight line, what can be said about the uptake of oxygen by the organisms in the respirometer?*

**b** *How much oxygen was absorbed by the organisms? State the amount as millimetres cubed per hour per milligram ($mm^3 hr^{-1} mg^{-1}$) of living material.*

**c** *How does the rise in temperature affect the rate of oxygen uptake? How can the effect be best expressed?*

**d** *What rate of oxygen uptake would you expect if you raised the temperature to 40° or 50 °C? Give your reasons.*

## INVESTIGATION
## 5F Respiratory quotient

(*Study guide* 5.7 'Cellular respiration and the mitochondrion'.)

It is possible, using the same apparatus as in investigation 5E, to discover something about the nature of the material being respired by a tissue or organism. This is done by comparing the amount of carbon dioxide given off with the amount of oxygen absorbed by the same tissue in the same time. Such a comparison is simply expressed as:

$$\text{Respiratory quotient (R.Q.)} = \frac{\text{volume of carbon dioxide given out}}{\text{volume of oxygen absorbed}}$$

These volumes must be measured in the same units for the same period of time, keeping the temperature and other conditions constant.

The respiratory quotient is useful because its value depends on the type of compound being respired. Carbohydrates such as glucose give a respiratory quotient of 1.0, while a fat such as tristearin has an R.Q. of 0.69. Examples of respiratory quotients for other compounds are:

| Amino acid glycine | 1.33 | Citric acid | 1.33 |
| Amino acid leucine | 0.80 | Oxalic acid | 4.00 |
| Amino acid lysine | 0.86 | | |

Therefore, if we measure the R.Q. of a tissue or organism this may suggest what type of compound is being respired.

As before, the volume of oxygen absorbed in a given time is obtained by measuring the change in gas volume in the presence of potassium hydroxide. The carbon dioxide produced is then obtained by measuring the volume change over the same period of time without potassium hydroxide present (see step **4** below).

*Procedure*

1  Using a respirometer, find the amount of oxygen absorbed (Vol. 1) by germinating seeds in a period of 45 minutes at 20 °C. Follow the same procedure as in investigation 5E; adjust the syringe to keep the manometer fluid at a constant level.

2  Remove the potassium hydroxide solution from both vessels and wash them out with water.

3  Replace the basket containing the germinating seeds in one vessel, an equivalent volume of water in the other vessel, and the bungs in both. Set the respirometer in the water bath at 20 °C and record any increase or decrease in gas volume over the next 45 minutes (Vol. 2).

4  Calculate the volume of carbon dioxide produced as follows:

Vol. 1 = oxygen absorbed $(cm^3)$
Vol. 2 = carbon dioxide produced − oxygen absorbed $(cm^3)$
Therefore, carbon dioxide absorbed = Vol. 1 + Vol. 2 $cm^3$,

and $\quad R.Q. = \dfrac{Vol. 1 + Vol. 2}{Vol. 1}$

*Questions*

a  *Is it possible to deduce from the respiratory quotient what substance, or type of compound, has been respired by the seeds?*

b  *What respiratory substrate is most likely to be present in the seeds, and what tests could you perform to discover the nature of the substrate? Does the R.Q. correspond with your expectations?*

c  *What would you expect to be the R.Q. of a growing culture of yeast?*

d  *Are question c and its answer in any way relevant to the respiration of germinating seeds? Give your reasons.*

e  *Suggest three different hypotheses to account for a respiratory quotient of unity (that is, 1.00).*

# CHAPTER 6

# HETEROTROPHIC NUTRITION

**INVESTIGATION**

**6A Digestion by micro-organisms and tissues**

(*Study guide* 6.4 'Digestion'.)

Animals and many micro-organisms rely on complex chemicals, taken in as food, to provide them with energy and nutrients. This type of feeding is termed *heterotrophic*. Organisms which, by contrast, need only simple chemicals are known as *autotrophic* feeders.

Heterotrophic organisms almost always modify their food substances by digestion before absorbing them. This can be investigated by placing a piece of gut from an animal, or some micro-organisms, in contact with starch. This is a very common food substance and its presence is easy to detect. If the starch is mixed in with agar jelly it can be handled easily.

*Procedure*

1   You will need two Petri dishes (with lids) each containing a suspension of starch in agar.

2   Place organisms and tissues on the surface of the agar or in wells cut in the agar with a small cork borer (*figure 10*). Do not put more than four items on any one agar plate. Some possibilities are:

A drop of yeast suspension.

A small portion of the gut of a locust (see investigation 6B for details of how to dissect out a locust gut).

Half a germinating grain of barley.

A drop of amylase solution.

A drop of human saliva.

A drop of water. (This is a control. Should it be sterilized, distilled, or tap water?)

Decide whether each specimen should be in a well or on the surface. In either case be sure the specimen makes good contact with the agar.

3   Label each dish carefully so that you can keep a record of the contents. Place the dishes in an incubator for 48 hours at 25 °C.

4   Remove the dishes from the incubator and take off the lids. Pour a solution of iodine dissolved in potassium iodide liberally over the surface of the agar in each dish. Wait 15–30 seconds and then pour it away.

5   Examine the plates carefully, holding each in turn in front of a lamp. Record patches and zones of different colour. Place the used plates in a refrigerator for future reference.

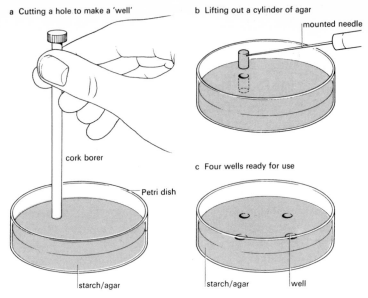

a Cutting a hole to make a 'well'

b Lifting out a cylinder of agar

mounted needle

cork borer

c Four wells ready for use

Petri dish

starch/agar

starch/agar     well

**Figure 10**
Preparing a starch/agar plate for the addition of organisms and tissues.

6  Before interpreting the results consult *Study guide I*, page 175, or any textbook which describes the structure of starch molecules.

*Questions*

a  *Unchanged starch forms a deep blue colour with iodine solution. Which organisms or other items have a colourless zone around them?*

b  *Are such zones in the agar only very close to the items or do they extend well away from them?*

c  *What can you say about the chemical composition of the clear zones of agar? What further tests could you perform on these plates to confirm your ideas?*

d  *What other colours have you observed besides blue? (Look particularly at the boundary between the blue and clear zones.) How could you account for other colours?*

e  *In the light of your answers to questions a, b, and c, relate this investigation to the process of digestion. Is there any significance in the similarity of the effect of animal and plant tissue and micro-organisms on starch?*

## INVESTIGATION

### 6B Digestive organs: a model gut

(*Study guide* 6.5 'The double function of the alimentary canal'.)

In the previous investigation you found that a wide variety of organisms will modify a substance such as starch before absorbing and utilizing it. One reason for this can be examined with the help of a model gut. This is constructed from Visking tubing, representing the gut and containing some food (starch). The digestive enzymes are provided by placing cut portions of gut from a freshly killed animal in the tubing.

*Procedure*

1 Take a freshly killed adult locust and place it in its normal resting position in the centre of a shallow dish which has a layer of hard wax at the bottom. Fix it with pins through the legs and thorax. Cut off the wings if they get in the way.

2 Pull back the tip of the abdomen with forceps, stretching it slightly, and insert a pin to hold it back. With fine pointed scissors pierce the body covering (exoskeleton) between two segments at about the position shown in *figure 11*. Cut down between the segments on one side and then continue the incision forward to the head along the dotted line shown. As you cut, let the lower blade of your scissors pull the locust up slightly. In this way you will be less likely to damage internal organs.

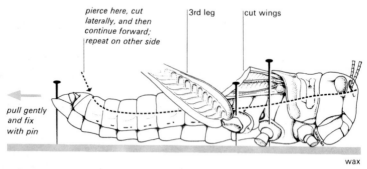

**Figure 11**
Side view of a locust prepared for dissection.

3 Make a similar cut on the other side and carefully remove the strip of exoskeleton from the animal so that you can see the internal organs. The most prominent of these is the gut, or alimentary canal (see *figure 12*). Remove the gut and place it in a watch-glass.

4 Cut the gut into small pieces with scissors, put it into a second watch-glass and cover it with a little starch suspension (1 per cent).

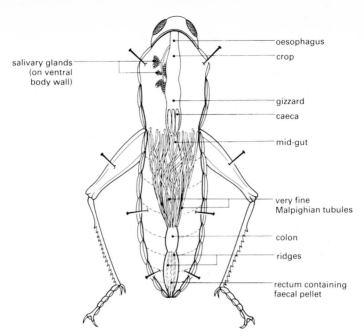

oesophagus
crop

salivary glands
(on ventral
body wall)

gizzard
caeca

mid-gut

very fine
Malpighian tubules

colon

ridges

rectum containing
faecal pellet

**Figure 12**
A dissected locust showing the alimentary canal.
*After Thomas, J. G.,* Dissection of the locust, *Witherby, 1963.*

5  Cut the tissue into smaller pieces using mounted needles or pointed scalpels.

6  Tie a knot in the free end of a roll of Visking tubing and cut it 15 cm from the knot.

7  Pour the cut-up tissue into the tubing and add more starch suspension until the total volume is about $10 \, cm^3$. Mix well by squeezing the tube between finger and thumb.

8  Lower the Visking tubing into a test-tube until the knot is close to the bottom. Use a bulldog clip or paper clip to make sure that the open end does not fall right in (see *figure 13*). Add distilled water on the outside of the tubing just up to the level of the liquid inside.

9  Set up a similar arrangement of Visking tubing, this time containing only $10 \, cm^3$ of starch suspension, in a second test-tube. Add distilled water on the outside of the tubing as before.

10  Incubate both tubes in a water bath at $25-30 \,°C$ for 25–30 minutes.

11  Remove a drop of liquid from the gut/starch mixture and add one drop of iodine solution on a white tile. Take a further $1-2 \, cm^3$ of the gut/starch mixture and add it to an equal volume of Benedict's solution in a test-tube. Heat the mixture until it boils. Record any changes of colour resulting from these tests.

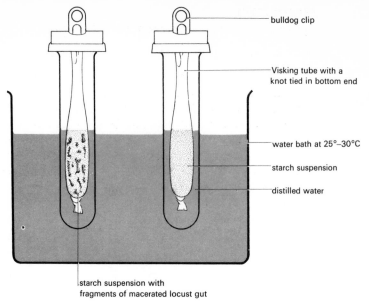

**bulldog clip**

**Visking tube with a knot tied in bottom end**

**water bath at 25°–30°C**

**starch suspension**

**distilled water**

**starch suspension with fragments of macerated locust gut**

**Figure 13**
Using Visking tubing as a model gut.

**12** Perform the same tests on: the water surrounding the tube containing the gut/starch mixture; the contents of the Visking tube containing starch suspension only; the water surrounding this Visking tube. Wash your pipette carefully after taking each sample to avoid accidental mixing of the test materials. Record your observations.

*Questions*

**a** *Iodine forms a blue compound with starch; Benedict's solution forms an orange precipitate when heated with reducing sugars. In which samples did you find starch and in which did you find reducing sugars?*

**b** *From these observations alone, is there any evidence that gut tissue changes starch to reducing sugars? Explain your answer.*

**c** *How could you use the starch/agar plates from investigation 6A to add to this evidence?*

**d** *Consider the answers to question a. What do these observations and the properties of Visking tubing suggest to you about the problem of digestion for a living locust feeding on starch-filled leaves?*

## INVESTIGATION

**6C   Digestion and absorption in the gut of a mammal**

(*Study guide* 6.5 'The double function of the alimentary canal'.)

Oxygen in the air is acceptable to organisms as it stands and can be absorbed without any preliminary chemical change. This is not so with food. An alimentary canal has two functions: digestion, to convert nutrients into substances which can be absorbed; and the process of absorption itself. In this and the following investigation the structure of the mammalian gut is examined to see how it is related to these functions.

If you can dissect a small mammal very soon after it has been killed you may be able to observe movements of the gut. These will be the peristaltic movements normal in the living animal. You will also be able to analyse, by chromatography, the contents of the various sections of the alimentary canal. From this it should be possible to deduce in which sections there has been significant digestion of recently eaten food.

*Procedure*

1   Heat approximately $1\,dm^3$ of Ringer's solution in a thermostatically controlled water bath to a temperature of 36–38 °C.
2   As soon as the mammal has been killed, place it on its back in a wax-bottomed dish and rapidly fix it in position with pins.
3   Cover the animal immediately with warm Ringer's solution.
4   Working as swiftly as possible, take hold of the skin in the middle of the abdomen with a pair of forceps, pull it upwards, and cut a hole in the skin with a pair of scissors (see *figure 14*). From this hole cut forwards with the scissors right up to the jaw and backwards towards the tail. Separate the skin from the muscle underneath and pin it back.
5   Carefully make a hole in the muscular wall beneath the skin at about the same point as before. Cut forward through the muscle until you come to the flap of cartilage (the xiphisternum) at the base of the sternum. Cut along the line of the lowest ribs, left and right, as shown in *figure 14*. Make similar cuts nearer the tail and fold back the muscle wall to expose the contents of the abdomen.
6   As soon as you have pulled back the skin and flaps of muscle out of the way, stop dissecting and carefully observe the contents of the abdomen. Look for movement of any kind, using a binocular microscope.
7   Check the temperature and replace the Ringer's solution if necessary.
8   If you see movements, record your observations carefully and then proceed to use *figure 15* to help you to identify the main parts of the alimentary canal. This exploration may start more intestinal

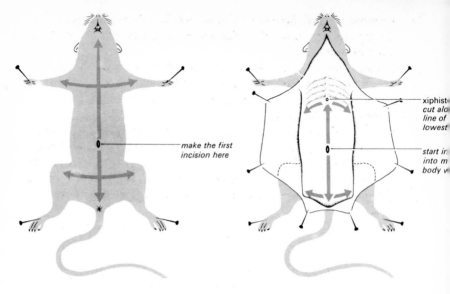

**Figure 14**
Dissecting the mouse – opening the abdomen.

movements, in which case stop and watch them. There are at least three distinct patterns of movement known to occur in the gut.

9   Remove the stomach, duodenum, ileum, caecum, colon, and rectum separately. Wash out the contents of each part with a few drops of warm Ringer's solution into separate labelled specimen tubes. Keep the volume of Ringer's solution to a minimum to obtain a concentrated extract.

10  Return the parts of the gut to warm Ringer's solution and cut out small sections from the walls of each.

**Figure 15**
The alimentary canal of the mouse.
*After Rowett, H. G. Q.*, Dissection guides 3: The rat, with notes on the mouse, *2nd edn, John Murray, 1952.*

11 Examine each of these in turn, mounted in a drop of Ringer's solution under a binocular microscope, to discover the form of the internal surface.

12 Record your observations, as far as possible, by simple sketches.

13 Add 0.5 cm³ distilled water to the contents of the specimen tubes, shake well, and leave to settle. Use the clear liquid for chromatographic analysis.

14 Prepare a thin layer plate by drawing a faint pencil line 2 cm from one end as in *figure 16*. Mark two small crosses on the line 7 mm from each edge. The crosses must not be any closer to the edge than this because the edge of the plate affects the smooth running of the chromatogram. Place some of the clear liquid from the stomach

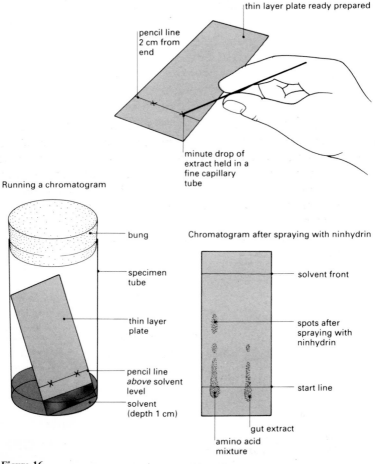

**Figure 16**
Producing a thin layer chromatogram.

extract on to one of these crosses. To do this dip a clean capillary tube into the solution and apply a small drop to the plate, using a quick delicate touch. Practise on a small piece of filter paper until you have produced spots not more than 0.5 cm in diameter. Place successive drops of extract on the same spot on the plate, allowing each to dry before adding the next. In this way a concentrated spot of extract, no wider than 5 mm, can be built up.

15 Use the same technique to load the solution of pure amino acids on the other pencil cross. One or two drops of this solution should be sufficient.

16 Use two separate labelled plates for the analysis of the contents of the duodenum and ileum (you may analyse the contents of the other parts of the gut too if you have time). When the spots are dry, put each plate in a specimen tube as shown, containing solvent to a depth of 1 cm. The solvent is a mixture of butan-1-ol, glacial ethanoic acid, and water in the proportions 80:20:30. Make sure that the spots are above the surface of the solvent. Stopper the tubes.

17 After one hour, or when the solvent front has reached the top of the plate, take the plate to a fume cupboard. Dry it and spray thinly with ninhydrin solution. This *MUST* be done in a fume cupboard with the extractor fan on or out of doors well away from other people; it must *NEVER* be done in the laboratory.

18 Heat the sprayed plate in an oven to 110 °C for about ten minutes.

19 Examine each plate carefully for spots of purple and other colours. These colours will fade with time, so make a record of the results.

*Questions*

a *Which parts of the alimentary canal appeared to move and how do you account for the movement?*

b *It is common knowledge that food passes along the alimentary tract from mouth to anus. Do the gut movements appear to move food in this direction? If not, what do the movements appear to do to the food inside the gut?*

c *Does the inside surface of the gut appear to be adapted to a digestive or an absorptive function, or both? Apply the question to each part of the gut that you examined, and state the reasons for your conclusions.*

d *The solution rising up the thin layer plate separates various products from the digestion of proteins. Ninhydrin forms coloured compounds with amino acids. From your observations, is there any evidence that proteins have been digested?*

**e** *Describe how the chromatogram of the gut extract compares with that of the mixture of pure amino acids. Explain the differences which you have observed.*

**f** *Which parts of the gut appear to be chiefly responsible for absorbing digested food? (Confirm your answer by reference to a textbook.) It would be reasonable to assume that absorption occurs in the last section of the gut, when the food is completing its journey through the alimentary tract. How do you account for the fact that this is not so?*

## INVESTIGATION
## 6D The fine structure of the intestinal wall

(*Study guide* 6.7 'Absorption'.)

After studying the structure and function of the major segments of the alimentary canal we now examine one of these segments, the ileum, in microscopic detail. The ileum is chosen because it is possible to see how it is adapted to perform both digestion and absorption. Digestion itself involves two processes: the action of enzymic secretions, and the thorough mixing and movement of the contents of the gut. Look, therefore, for the characteristic structures of glands and secretory cells, as well as the muscle fibres responsible for peristalsis and other movements. Look, also, for structural features you might expect to find in any organ adapted to carry out efficient absorption.

*Procedure*

1  Use a prepared, transverse section of the ileum of a rat or mouse. Note the staining technique employed, as indicated on the label. If possible compare sections that have been stained with different reagents, for example P.A.S. (periodic acid, Schiff method) and H.E. (haematoxylin and eosin).

2  Examine the section with the naked eye and with a hand lens ($\times 10$) or L.P. microscope. Compare the appearance with sketches made of samples of gut from the dissected animal.

3  Using either the width of the field of view in your microscope on L.P., or an eye-piece graticule, estimate the length of a typical villus in your section of ileum.

4  Work out how the two-dimensional section you are examining is related to the three-dimensional structure of the ileum.

5  Look at the section under higher power magnification ($\times 100$, $\times 400$).

6  Use both L.P. and H.P. to identify on the sections the structures indicated in *figure 17*. At the same time, answer the questions.

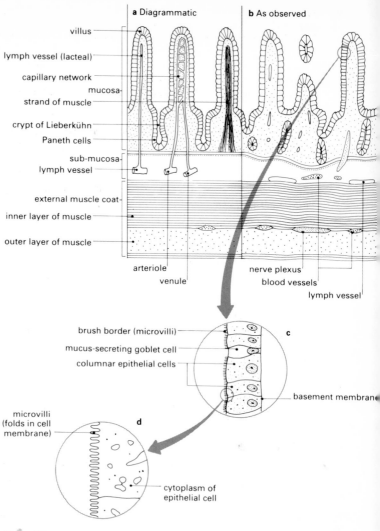

a Diagrammatic   b As observed

villus
lymph vessel (lacteal)
capillary network
mucosa
strand of muscle
crypt of Lieberkühn
Paneth cells
sub-mucosa
lymph vessel
external muscle coat
inner layer of muscle
outer layer of muscle

arteriole
venule
nerve plexus
blood vessels
lymph vessel

brush border (microvilli)
mucus-secreting goblet cell
columnar epithelial cells
c
basement membrane

microvilli
(folds in cell
membrane)
d
cytoplasm of
epithelial cell

**Figure 17**
The microscopic structure of the wall of the ileum.
**a** A highly schematic drawing showing the structures within the wall of the ileum.
(Magnification approximately ×40.)
**b** More realistic view of how these structures appear in a section.
**c** Part of the epithelium which lines the ileum magnified approximately ×200.
**d** Part of an epithelial cell magnified approximately ×1500.
*Based on Roberts, M. B. V., Biology, a functional approach, Students' manual, 1st edn,
Nelson, 1974.*

*Questions*

a  *What is the approximate length of a typical villus in your section of the ileum?*

b  *Explain why many villi appear in the section as islands of tissue with no connection to the wall of the ileum. (See* figure 17b.*)*

c  *Make a simple sketch of your section to record the distinct layers of tissue in the wall of the ileum. Indicate on your sketch the arrangement of muscle fibres in the two layers of muscle.*

d  *Examine the orientation of muscle fibres in the inner and outer layers of muscle. Which layer possesses longitudinal muscle fibres? What effect will the contraction of the inner and outer layers of muscle have on movements of the intestine?*

e  *Does the diagram in* figure 17a *and* b *represent a longitudinal or transverse section of ileum? Explain your answer.*

f  *If your section is stained by P.A.S., mucus and cells producing it (goblet cells) will show up clearly, stained deep pink. Comment on the frequency and distribution of the goblet cells. What part do you think mucus plays in the action of the alimentary canal?*

g  *Can you identify any other secretory cells? If so, record their position on a simple, outline sketch.*

h  *Where in the wall of the intestine are blood vessels situated? Refer to* figure 18, *which is a photomicrograph of an injected specimen. Through what structures must digestive products pass to reach the blood stream?*

i  *Look at* Study guide *figure 153b, an electronmicrograph of a minute portion of the inner surface of a human ileum. Estimate the length of the microvilli in this figure from the magnification ( × 40 000) of the electronmicrograph. Explain the relationship of the epithelial cells which line the ileum to:*
*1   the villi, and 2 the microvilli.*
*What is the significance of the villi and microvilli for the function of the ileum?*

j  *Mannose and glucose are absorbed across the wall of the ileum at very different rates. Glucose passes five times more rapidly than mannose. Yet they are both hexoses ($C_6$ sugars) and are therefore very similar compounds. What kind of biological mechanism could distinguish between such similar molecules? Where in the wall of the ileum might this mechanism be located?*

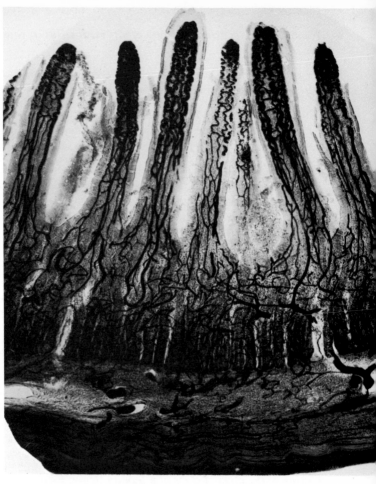

**Figure 18**
A photomicrograph of the ileum of a cat with blood vessels injected ( × 60).
*From Freeman, W. H. and Bracegirdle, B.,* An atlas of histology, *2nd edn, Heinemann 1967.*

## INVESTIGATION
## 6E  A microscopic investigation of the liver

(*Study guide* 6.8 'The role of the liver'.)

The liver is a 'control centre' and 'sorting house' for metabolism in t
vertebrate animal. This general role includes a great variety of functio
ranging from assimilation of the products of digestion and the control

blood composition to a number of very specialized functions, such as the production of bile.

Many of the substances absorbed from food are carried in the hepatic portal vein directly to the liver. Here the hepatocytes (liver cells) remove several of these compounds from the blood and process them according to the needs of the body. For instance, glucose can be used in several ways: (a) it may be converted into glycogen for storage in the liver; (b) it may be converted into fatty acid for storage in fat tissue; (c) it may be used to synthesize amino acids; (d) it may be respired to provide energy for the hepatocyte.

It would be wrong to regard the liver as nothing more than a 'clearing house' for incoming nutrients. Cells throughout the body are continually generating waste products, such as unwanted amino acids, and ammonia. The hepatocytes convert these compounds into relatively harmless products, such as urea, which can be excreted safely.

Under the influence of hormones, hepatocytes control the rate at which sugars, amino acids, and fats are removed from the blood. They also regulate the rate at which these compounds are released back into the blood. In this way the liver influences the chemical composition of blood and, in turn, the composition of every other body fluid. The hepatocyte plays a central role in the maintenance of homeostasis in vertebrates.

The liver also has a number of specialized functions such as the production of bile, plasma proteins, and cholesterol. A general textbook should be consulted for a complete description of these functions.

Hepatocytes are arranged in a highly organized way within the liver. This is necessary because each hepatocyte must communicate with three different blood vessels and the bile duct. Blood arrives in the liver through the hepatic artery and the hepatic portal vein. Blood and bile are conducted away from the liver in the hepatic vein and the bile duct, respectively. The aim of this investigation is to examine the microscopic organization of the liver in order to find out how each hepatocyte communicates with the rest of the body.

*Procedure*

1  Examine under L.P. a prepared section of liver that has been stained with haematoxylin and eosin. Haematoxylin is a basic dye which stains structures such as the nucleus and ribosomes blue or purple. Eosin is an acidic dye which stains proteins in the cytoplasm red. (You may need to switch to H.P. to see these colours clearly.)

2  Again turn to L.P. on the microscope, or better still, use a hand lens to examine the section of liver. Note how the tissue is organized. It should appear to be divided into structures which are very roughly

hexagonal in cross section. These are the *lobules* of the liver. In a section of pig's liver it is usually clear that each lobule is separated from its neighbours. This is less obvious in human or rat liver. Try to imagine what a group of lobules will look like in three dimension. *Figure 19* will help.

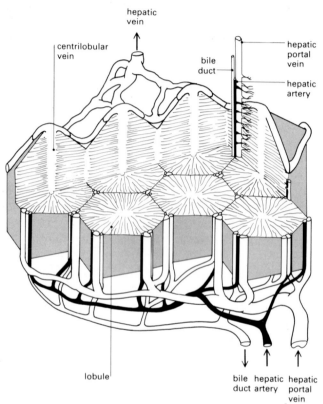

**Figure 19**

The overall arrangement of the liver lobules, the vascular system, and the bile collecting system. Branches from the hepatic portal vein and the hepatic artery run along *portal tracts* between the lobules. Blood from both supplies passes among the hepatocytes through spaces called *sinusoids* which converge to drain into a central vein. These merge together to form the hepatic vein.

In contrast, bile is secreted into a network of minute *bile canaliculi* which pass around the hepatocytes. The canaliculi are too small to be represented in this diagram. They drain into a system of collecting ducts which also run in the portal tracts but in the opposite direction to the two blood supply systems. These ducts drain ultimately into the common bile duct with its gall bladder.

*Based on Wheater, P. R., Burkitt, H. G. and Lancaster, P. Colour atlas of histology, Longman, 1985.*

3  Measure the diameter of a typical lobule using an eyepiece graticule and the low power objective.

4  Occasionally in the angle between two or three adjacent lobules you will find a group of vessels cut in cross-section. This is called a *portal tract*. Examine one under low and high power and make a large drawing of it with its neighbouring lobules. By examining the walls of the vessels and by comparing them with the structures in *figure 20* decide what each one is.

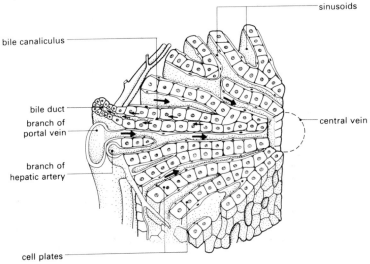

**Figure 20**
Diagrammatic representation of the arrangement of hepatocytes into plates radiating out from the central vein. The spaces, or sinusoids, between the cell plates connect with the blood vessels of the portal tracts and with the central vein. The large arrows indicate the direction of the flow of blood, the small ones, bile.
*Based on Bloom, W., and Fawcett, D. B., A textbook of histology, 10th edn, W. B. Saunders, 1975.*

5  Look for a blood vessel in the centre of a lobule. Examine its wall and decide whether it is a vein or an artery.

6  Note the arrangement of hepatocytes in rows radiating outwards from the central blood vessel. Examine a row under high power. Look for any cells that may line the sinusoid, or space, between two adjacent rows of hepatocytes. Measure the width of a sinusoid using an eyepiece graticule and high power.

*Questions*

a   *How many vessels have you distinguished in the portal tracts? What are they, and how do they differ from each other?*

b   *Describe the arrangement of hepatocytes inside a lobule. Record the diameter of a lobule and the width of a sinusoid.*

c   *What fills the sinusoids between the rows of hepatocytes in the living animal? As it is not possible to examine the living tissue, you could only deduce the answer by examining the structures that connect with a sinusoid.*

d   *Does blood flow from the outside of a lobule inwards, towards the central blood vessel, or in the opposite direction? Explain how you arrive at your answer.*

e   *List the structures or vessels that blood must pass through in travelling from the hepatic portal vein to the hepatic vein.*

f   *From your observations and by examining figure 21, can you tell whether the hepatocytes and the blood are in direct contact or are kept separate from one another?*

g   *What evidence can you obtain from examining the prepared section and the electronmicrograph in figure 21, to suggest that the metabolism of the hepatocyte is very active?*

h   *Is there any evidence from the appearance of hepatocytes to suggest that these cells might be grouped into different kinds, according to the functions which they carry out?*

i   *Can you find any evidence on your slide to suggest how bile that has been secreted from a hepatocyte could be channelled into a bile duct?*

j   *List the ways in which the composition of blood in the hepatic vein might differ from that of blood in*
   *1   the hepatic portal vein during digestion and absorption of a meal, and*
   *2   the hepatic artery.*

k   *An examination of the microscopic structure of the liver helps us to understand how the liver works. Despite this, it does not tell us what the liver actually does. What type of investigation would be most useful in establishing the functions of the liver?*

microvilli | | bile canaliculus

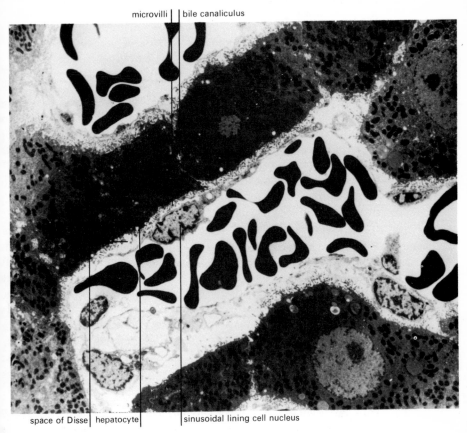

space of Disse | hepatocyte | | sinusoidal lining cell nucleus

**Figure 21**

Electronmicrograph (×3200) showing the main ultrastructural features of the liver.
Each hepatocyte is bathed on at least two sides by blood. The sinusoids are lined by a
discontinuous layer of cells. There is a space, the *space of Disse*, between these
sinusoid lining cells and the neighbouring hepatocytes. The space of Disse is continuous
with the sinusoid lumen. Numerous microvilli extend from the hepatocyte surface into
the space of Disse, greatly increasing the surface area of the plasma membrane.

The hepatocyte is crowded with organelles, particularly rough and smooth
endoplasmic reticulum, mitochondria, and lysozomes. Lipid droplets and glycogen
rosettes are present in variable numbers depending on nutritional status.

Bile canaliculi are seen to be formed from the plasma membranes of adjacent
hepatocytes. Small microvilli project into the canaliculi.

*From Wheater, P. R., Burkitt, H. G., and Daniels, V. G.,* Functional histology – a text
and colour atlas, *Churchill Livingstone, 1979.*

# CHAPTER 7 PHOTOSYNTHESIS

**INVESTIGATION**
## 7A The interaction of plants and animals

(*Study guide* 7.1 'Photosynthesis and the biosphere'.)

So far our approach to biology has been to investigate organisms individually and their processes, such as digestion or respiration separately. Normally, many processes function together in a living organism. For instance, when a plant is exposed to light we should expect it to photosynthesize and respire at the same time. The overall effect of the plant's gaseous exchange on the composition of its surroundings will result from the combination of these opposing processes.

In their natural surroundings organisms are not isolated from others. It is usually very difficult to unravel the effect that a single organism has on its environment when it is part of a complex community. To simplify this problem we can set up artificial communities in the laboratory with just a small number of species. In this investigation simple terrestrial and aquatic 'communities' are established. With the terrestrial community you can study how animals affect the level of carbon dioxide in the atmosphere surrounding plants. The aquatic 'communities' are used to determine the relative biomass of plants and animals that can live together while maintaining a balanced level of $CO_2$ in their surroundings.

*Procedure*
Note: It should be possible to set up the terrestrial community (steps **1**–below) and the first set of microaquaria (steps **10**–**14** below) at the same time. Both experiments can be left and examined twenty-four hours later.

*The compensation period in a terrestrial community*
1  Consider questions **a**–**d** as you set up your apparatus.
2  Select two containers large enough to contain several plants. The containers should be transparent, to allow the entry of light, but easy to seal, to prevent the entry or escape of gases. See *figure 22.*
3  Choose a number of plant shoots, all from one species, that can be stood in pots of water in the containers. Place equal quantities of shoots in each container.
4  Add small animals such as woodlice to one container; their total mass should be between 5 % and 10 % of the fresh mass of the plants.
5  Bubble air through a solution of hydrogen carbonate indicator until this is red. Devise a means of exposing samples of this indicator at

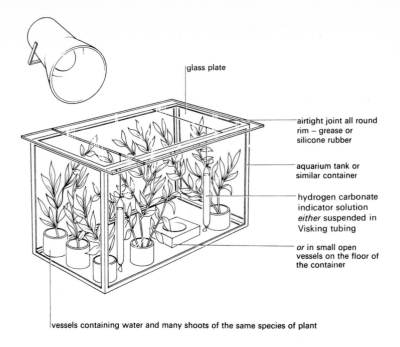

glass plate

airtight joint all round rim – grease or silicone rubber

aquarium tank or similar container

hydrogen carbonate indicator solution *either* suspended in Visking tubing

*or* in small open vessels on the floor of the container

vessels containing water and many shoots of the same species of plant

Two methods of supporting hydrogen carbonate indicator

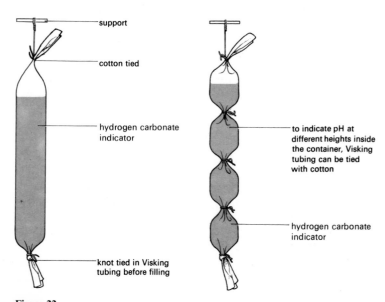

support

cotton tied

hydrogen carbonate indicator

knot tied in Visking tubing before filling

to indicate pH at different heights inside the container, Visking tubing can be tied with cotton

hydrogen carbonate indicator

**Figure 22**
An example of apparatus used to find the compensation period of plants.

various places inside the containers. One convenient method is shown in *figure 22*. Visking (cellulose) tubing will hold liquid yet allow the penetration of surrounding gases. The solution inside will not respond to changes in external gas composition as quickly as an indicator in open dishes, but these would have to be on the floor of the container whereas the Visking tubes can be suspended at any height. Seal the containers, which must be airtight.

6   Set aside a sample of hydrogen carbonate indicator in a sealed glass container so that it may be used later for comparing colour.

7   Ideally, the plants in their containers should be placed in the dark for about twelve hours. This is best done by putting them in a light-proof cupboard fitted with a lamp and time-switch. The latter should be set to switch the light off twelve hours before the next laboratory session. If a time-switch is not available it is best to cover the containers with an opaque box or cloth at the end of a school day. This will give a longer dark period than desirable, *e.g.* 5 p.m. to 9 a.m., but will nonetheless give useful results.

8   After the dark period, note the yellow colour of the indicator and illuminate the containers equally. Record the power of the light source and measure the distance between it and the containers so that the experiment can be repeated exactly, if necessary.

9   Record the times taken for the yellow hydrogen carbonate indicator to return to its original red colour, in both containers. These times are known as *compensation periods*, and the stage at which carbon dioxide output is balanced by uptake is the *compensation point*.

*Balance within a microaquarium*

10   Set up microaquaria (see *figure 23*). You can use large, corked test-tubes containing hydrogen carbonate indicator, on its own, with pond weed, with aquatic animals, and with both pond weed and animals.

11   Place the tubes at equal distances from a light source and leave them for 24 hours. You may need a shield to avoid over-heating. *AVOID* looking directly at the light source: it could damage your eyes.

12   Compare the colours of the indicator in the four microaquaria. Note whether the colour of the fourth is nearer that of the indicator containing animals only, or plants only.

13   Using this observation (**12**) set up a further series of microaquaria containing plants and animals in proportions which produce no colour change during 24 hours of illumination.

14   Remove the contents of the most 'balanced' microaquaria and weigh the animals and plants.

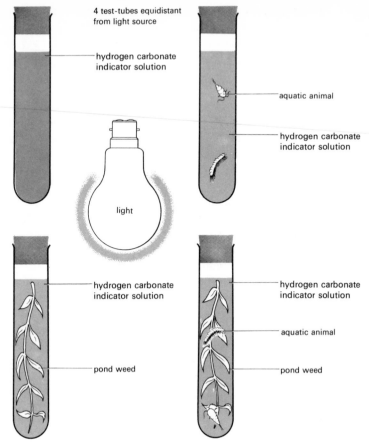

**Figure 23**
Microaquaria.

*Questions*

a By what means can you ensure that the quantities of plants in both containers are equal (step 3)?

b Is it better to stand the plants in soil or water (step 3)? Give your reasons.

c Should steps be taken to maintain a constant temperature in the containers and if so, why?

d The change in colour of the hydrogen carbonate indicator is gradual. It is difficult to assess accurately the time it takes for it to become a certain shade of red again. What steps can you take to make the measurements more precise?

e   *Is the size of the container important in relation to the number of plants used? Would you expect different results using large containers with a few plants?*

f   *What is the proportion of animals to plants used in step 4 as determined by fresh mass?*

g   *Was the indicator restored to its red colour when the container was illuminated (step 9)? From this observation state whether the community contained too many animals for continued survival or whether it could accept more.*

h   *What proportions of animals and plants, by mass, form a balanced microaquarium community in terms of carbon dioxide concentration (step 14)?*

i   *Suppose that you wanted to set up a balanced aquarium community. Would steps 10 to 14 above form a useful method for determining the proportion of animals and plants to be used?*

## INVESTIGATION
## 7B   The structure of a leaf

(*Study guide* 7.2 'The sites of photosynthesis'.)

Leaves come in a great variety of shapes but, with few exceptions, they have certain features in common: they are thin, flat, and green. It seems reasonable to assume that these characteristics are related to the photosynthetic functions of leaves. In order to synthesize carbohydrates leaves absorb light, take up water, and engage in gaseous exchange. A microscopic examination of their internal structure reveals something about how they perform these functions.

In this investigation the detailed structure of leaves from several different types of plant is examined. Thin transverse sections of leaves are very difficult to cut without considerable practice or a microtome. Although you may not have to cut sections yourself, *figure 24* shows three techniques which are commonly used. *If you try these, take GREAT CARE to avoid cutting your fingers.* You will need to examine prepared slides of transverse sections. If possible these should include examples of leaves from a monocotyledon, a dicotyledon, and a xerophyte (a plant adapted to a dry environment). A leaf from a fresh specimen of *Elodea* can serve as an example from an aquatic plant. It is instructive to note which structural features are found in all these contrasting leaves. But it is also interesting to compare how in leaves adapted to totally different environments the basic plan is modified to suit their particular needs.

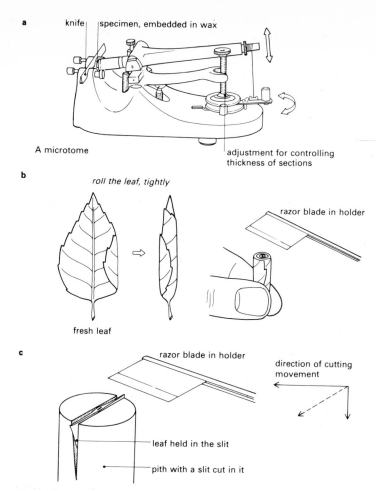

**a** knife ┃ specimen, embedded in wax

A microtome

adjustment for controlling thickness of sections

**b** roll the leaf, tightly

razor blade in holder

fresh leaf

**c** razor blade in holder

direction of cutting movement

leaf held in the slit

pith with a slit cut in it

**Figure 24**
Methods of cutting leaf sections. *TAKE CARE* when using the razor.

*Procedure*

1 Examine the transverse sections of leaves from a dicotyledon such as *Ligustrum* (privet) or *Syringa* (lilac) and a monocotyledon such as *Lilium* or *Zea mays* (maize). Examine each section first under low or medium power and then high power magnification.

2 Identify as many different types of cells as you can and for each section make a simple plan drawing to show the organization of these cells. Answer questions **a** and **b**.

3 Now examine the transverse section of a leaf from a xerophyte, such as *Ammophila arenaria* (marram grass). Again make a plan drawing of the internal organization. Answer question **c**.

**4**  Pick a leaf from a shoot of *Elodea* and mount it in a drop of water
under a coverslip. It may be curled and so easier to mount if
you first cut it into two or three pieces with fine scissors. Observe
under medium and high power magnification. Answer questions **d**
and **e**.

*Questions*

**a**  *Which of the structural features that you have observed are found:*
1  *in only the dicotyledonous leaf;*
2  *in only the monocotyledonous leaf;*
3  *in both types of leaf?*

**b**  *To what extent can you suggest explanations for the differences
between the structures of the monocotyledonous and dicotyledonous
leaves?*

**c**  *Which features of the leaf from the xerophyte contrast with the other
leaves which you have examined? Try to explain why the
xerophyte should possess such structures.*

**d**  *Describe the different cell types that you have found in the leaf of
Elodea.*

**e**  *A leaf of Elodea appears to be very simple when compared with one
from a typical 'dicot'. Explain how it can be an efficient
photosynthetic organ even though it lacks some of the structures
found in an aerial leaf.*

**f**  *What structural features of the leaves examined so far can be
regarded as adaptations to a photosynthetic function?*

## INVESTIGATION
## 7C  The evolution of oxygen

(*Study guide* 7.3 'The mechanism of photosynthesis'.)

A photosynthesizing leaf needs sources of light, carbon dioxide, and
water. A shortage of any of these will limit the rate of photosynthesis;
low temperature can also suppress photosynthesis. To find out which
these factors, in practice, limits photosynthesis we need a quantitative
method of measuring its rate.

The rate of production of oxygen by a plant might provide a suitable
measure of photosynthesis if it can be determined with ease. This
possible with aquatic plants such as *Elodea* which release bubbles
oxygen. Any simple apparatus for collecting these bubbles over a period
of time and measuring the volume of gas can be used. An example

shown in *figure 25*, though others may be devised. Indeed, useful results can be obtained simply by counting the rate of release of bubbles. However, because of variations in bubble size, this method is at best semi-quantitative and will not yield results with great precision.

For photosynthesis to occur there must be a source of carbon. For aquatic plants this is in the form of hydrogen carbonate ions ($HCO_3^-$) present in the water. Thus, when carrying out the investigation you should realize that this environmental factor may vary as well as light. An additional or alternative investigation may be carried out in which the light intensity is fixed but the concentration of hydrogen carbonate ions is varied. If temperature is varied, then allowance must be made for changes in the solubility of oxygen in water.

*Procedure*

1 Prepare an apparatus for collecting gas bubbles from a shoot of *Elodea*.

2 Put a bright light source near an aquarium tank containing *Elodea* for several hours. If no bubbles appear, add potassium hydrogen carbonate solution sparingly to the water.
*DO NOT look directly at the light source: it could damage your eyes. Shield the light to minimize risk.*

3 Choose a piece of *Elodea* which has a steady stream of bubbles coming from it. Transfer it to a test-tube or specimen tube of convenient size, filled with water from the tank. Set up the apparatus shown in *figure 25* (or a similar device for collecting gas).

4 Set the specimen tube in a larger vessel containing water and a thermometer so that you can check that the temperature remains constant through the following stages.

5 Put a bright light source such as a slide projector close to the apparatus so that light falls directly on the piece of pond weed. *Shield the other sides.* Darken the room to exclude other sources of light.

6 The easiest way to change the intensity of light falling on the plant is by moving the source away. Measure the initial distance between the *Elodea* shoot and the light source. (Be sure that you do not measure an arbitrary distance from, say, the edge of the beaker to the rim of the lamp holder!)

7 The volume of gas collected can be estimated (see *figure 25*) by drawing the bubble along the capillary until it is alongside the scale and measuring its length. Once measured, the bubble can then be drawn further along to the wide tube where it will remain conveniently out of the way during subsequent measurements. Measure the distance from the projector to the pond weed and the quantity of gas evolved in a known period, *e.g.* 5 to 10 minutes.

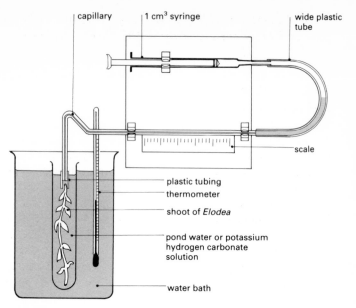

**Figure 25**
Apparatus for collecting and analysing gas from *Elodea*.

**8**   Double the distance between the light source and pond weed. Check that the temperature remains constant. Collect and measure second sample of gas in the same period as before.

**9**   Continue to make such measurements with the light source at greater and greater known distances from the pond weed.

**10**   The intensity of light falling on a given area, such as a piece of wee from a constant source, such as a projector, is inversely proportion to the square of the distance between them. If, for example, the distance is doubled, the light intensity is decreased not by 1/2 but $1/2^2$ or 1/4. Plot a graph of the amounts of gas collected in each equal period of time (vertical axis) against $1/d^2$ (horizontal axis) where $d$ is the distance between the plant and the light source.

*Questions*

**a**   *State any relation between gas production and light intensity which has been demonstrated by your results.*

**b**   *How would you confirm, experimentally, that the light intensity decreases with the square of the distance?*

**c**   *What would you expect the composition of the gas in the bubble to be? How would you determine this?*

**d** *What changes can occur in a bubble of gas as it rises from a piece of pondweed to the surface of the water? How might these changes affect your results?*

## INVESTIGATION
## 7D Leaf pigments

(*Study guide* 7.3 'The mechanism of photosynthesis'.)

Organisms which are capable of photosynthesis always possess pigments. It is likely, therefore, that these substances play a vital part in photosynthesis. Although the chlorophylls make most plants green, these are not the only photosynthetic pigments occurring in nature. Quite a large number of plants, such as copper beech and some seaweeds, are not green but are red or brown. It is quite possible that such tissues also possess the green chlorophylls but they are masked by other pigments. Similarly, green leaves may possess small amounts of non-green pigments. One way to discover whether this is so is to extract the pigments from leaves and to separate them from one another by chromatography. The success of this separation depends very much on the amount of care put into it. Alternative methods for preparing the extract are outlined below. Method (A) is relatively quick to complete but is unlikely to give the clean separation of pigments that is possible with method (B). The latter however requires more time, materials, and apparatus. Likewise, two types of chromatography are described. Thin layer chromatography is more rapid than paper chromatography and will handle much smaller quantities of material. Choose the methods and techniques according to what apparatus and time is available to you.

*Procedure*
As far as possible carry out the extraction of leaf pigments away from bright lights and keep the solutions cool. Make sure that your apparatus is as clean and dry as possible.

*Preparation of extract: Method A (About 1½ hours)*

1 Collect a handful of leaves from several different plants, some green, some red, and some variegated. Keep the collections separate.
2 Tear the leaves into small pieces and shred them in an electric grinder. (Alternatively the leaves can be ground in a pestle and mortar with acid-washed sand and a little propanone. *TAKE CARE: this is flammable.*)

After this point, carry out all procedures in a *FUME CUPBOARD*.

**3**  Scrape the leaf fragments into a large test-tube and add just enou
propanone to cover them. Shake well and allow the mixture to stand
at least an hour. Then decant off the clear coloured liquid. Use t
extract for chromatographic analysis. (This extraction can be speeded
considerably by placing the test-tube in a beaker of warm wat
However, the heat will also accelerate the degradation of the pigmen

*Preparation of extract: Method B (about 2½ hours)*

**4**  Perform steps **1–3** above. Filter the propanone extract from step **3** to
remove plant debris.

**5**  Fill a small funnel with filter paper and anhydrous sodium sulphate.
Pass the pigment solution through this to remove any water.

**6**  Pour some of the pigment solution into a Petri-dish or watch-glass.
Evaporate the solvent in a vacuum desiccator under reduced pressur
This takes about half an hour. (If a vacuum desiccator is not availabl

the solvent can be evaporated in a round-bottomed flask connected
directly to the vacuum pump.) *THERE MUST BE NO NAKED
FLAMES in the laboratory while this procedure is being carried out.*

**7**  If the residue does not appear quite dry add 2 cm$^3$ more of propanon
and evaporate down again.

**8**  Dissolve the dried residue of the plant pigments in a *few drops* of
propanone. Use this extract for chromatographic analysis.

*Paper chromatography (about 40 minutes)*

**9**  Prepare the solvent for chromatography fresh by adding one part of
per cent propanone to nine parts of petroleum ether (boiling range

80–100 °C) and pour it into a jar to a depth of 1 cm. Cover this with
lid. *TAKE CARE: these liquids are flammable.*

**10**  Cut strips of chromatographic paper about 5 to 6 cm by 20 to 30 cm.
The exact size will depend on its fitting inside the jar *so that its edges
do not touch the glass* (see *figure 26*). Handle the paper by the edges
because finger marks spoil the process of pigment separation.

**11**  Rule a light pencil line across a strip 2 cm from one end and put a
small drop of extract at its centre. Apply this extract with a fine
capillary tube or a micropipette. When this has dried put another dr
in the same place and repeat this process so that a small but
concentrated spot accumulates. It should not be wider than 3 to 4 m

**12**  After the spot has dried suspend the loaded paper vertically in the ja
with the spot at the bottom but *above* the surface of solvent (see *figu
26*). Close the container so that the paper is surrounded by air
saturated with solvent vapour.

**13**  The solvent will ascend the paper rapidly, carrying the pigment with
After 20 to 30 minutes remove the paper and allow it to dry in a fum
cupboard with the extractor fan on.

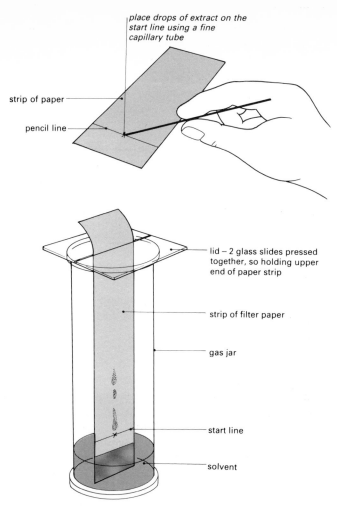

place drops of extract on the start line using a fine capillary tube

strip of paper

pencil line

lid – 2 glass slides pressed together, so holding upper end of paper strip

strip of filter paper

gas jar

start line

solvent

**Figure 26**
Setting up a paper chromatograph.

14  Repeat the procedure using the other extracts.
15  Count the number of spots on each paper. Describe the colours. As these fade quickly, outline each spot with pencil for future reference.
16  Though the distances between spots depends on duration and experimental conditions, the order from top to bottom is always the same. Refer to a textbook or other sources of information and try to name the pigments by colour and relative position.
17  The reference may use the term Rf. This is a quotient obtained by dividing the distance through which a substance has moved by the

distance through which the solvent has moved (in the same time and units). For example, if one component of a mixture rose 15 cm from the start line and the solvent moved 20 cm from the same line, then fo this component:

$$Rf = \frac{15}{20} = 0.75$$

Calculate the Rf values for each spot from one extract and see if these correspond with values given in a text or reference. (Bear in mind tha Rf values are affected by the methods of extraction and the solvents used.)

*Thin layer chromatography (about 30 minutes)*

18  Prepare the chromatography solvent as in step **9** above. Pour it into the specimen tube to a depth of 1 cm. Fit the bung into the tube.
19  If the thin layer plates are made on a plastic backing they can be trimmed carefully with scissors to fit inside the specimen tube.
20  Gently place a very small drop of the pigment solution onto the thin layer plate, 2 cm from one end. Follow the instructions in step **11** to apply more solution.
21  When the pigment spot is thoroughly dry stand the thin layer plate inside the specimen tube as in *figure 16* (investigation 6C, page 29). Make sure the pigment spot is just *above* the surface of the solvent. Replace the bung and leave until the solvent approaches the top edge of the plate (about 10 minutes).
22  Remove the plate and allow it to dry.
23  Repeat the procedure with the other extracts.
24  Analyse and record the results as suggested in steps **15**–**17** above.

*Questions*

a  *What is the essential difference between methods A and B of preparing the extract, and why does method B produce better results?*

b  *Do all the pigment extracts contain the same variety of component If not, give details.*

c  *Are some pigments common to all the extracts?*

d  *What conclusions can you draw from your answers to questions b and c concerning the role of pigments in photosynthesis? Try to fin out, from textbooks, what function is performed by plant pigmen*

## INVESTIGATION
## 7E  The reducing activity of chloroplasts: 'the Hill reaction'

(*Study guide* 7.3 'The mechanism of photosynthesis'.)

It is a relatively simple matter to show that illumination of a green leaf leads to the synthesis of starch or that this process needs a supply of carbon dioxide. However, it is more difficult to analyse the many chemical reactions that must occur between the absorption of light by chlorophyll and the enzyme-catalysed synthesis of starch. For instance, reduction of carbon dioxide, $CO_2$, to carbohydrate, $(CH_2O)_n$ could not take place without a source of reducing activity. R. Hill reasoned that if chloroplasts do produce a reducing agent it might be detected with a dye such as DCPIP (dichlorophenol-indophenol). This blue dye is readily reduced to a colourless compound by reducing agents.

*Procedure*
Between steps 1–6 all apparatus and solutions *must* be kept chilled.

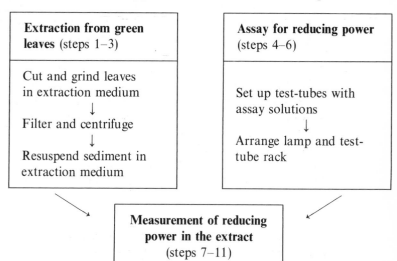

| **Extraction from green leaves** (steps 1–3) | **Assay for reducing power** (steps 4–6) |
|---|---|
| Cut and grind leaves in extraction medium ↓ Filter and centrifuge ↓ Resuspend sediment in extraction medium | Set up test-tubes with assay solutions ↓ Arrange lamp and test-tube rack |

**Measurement of reducing power in the extract** (steps 7–11)

### Extraction
1  Cut two or three green cabbage leaves into small pieces, removing the midrib. Thoroughly grind 20 g of the cut leaves in a pestle and mortar with 40 cm³ of sucrose/phosphate buffer ($0.3 \, \text{mol dm}^{-3}$ sucrose dissolved in pH 6.5 phosphate buffer).
2  Filter the extract through two layers of muslin or stocking material and pour the filtrate into centrifuge tubes.
3  Spin the extract in a bench centrifuge at high speed for ten minutes. Then pour off the liquid (supernatant) into a boiling tube and label

it 'supernatant'. Resuspend all of the green sediment in just $10 \text{ cm}^3$ c
sucrose/phosphate buffer; label it 'sediment'.

**Assay**

4   Set up the following assay solutions in five labelled test-tubes:

|  | Tube 1 | Tube 2 | Tube 3 | Tube 4 | Tu |
|---|---|---|---|---|---|
|  | Control<br>No extract<br>– | Control<br>No DCPIP<br>Sediment | –<br><br>Sediment | Control<br>No light<br>Sediment | –<br>Su<br>na |
| DCPIP<br>solution | 0.5 | – | 0.5 | 0.5 | 0.5 |
| Chloride/<br>phosphate buffer | 0.5 | 0.5 | 0.5 | 0.5 | 0.5 |
| Phosphate buffer | 4.0 | 3.5 | 3.0 | 3.0 | 3.0 |

All the volumes in this table are in $\text{cm}^3$.

5   Cover tube 4 in aluminium foil to exclude the light.
6   Arrange a strong (75–100 watt) lamp so that it points towards a
    test-tube rack. The distance between the rack and lamp should be
    12–15 cm.

**Measurement**

7   When the extract is ready add $1 \text{ cm}^3$ of the 'supernatant' to tube 5
    and $1 \text{ cm}^3$ of the 'sediment' to tubes 2, 3, and 4. Add nothing more
    to tube 1.
8   Shake each tube and note the time at which the contents were mixe
9   Place all five tubes in the illuminated rack and measure the time
    taken for the dye to lose its colour. Tube 2, a control without any
    dye, provides a reference for comparison with the other tubes.
10  If the extract is so active that it decolorizes the dye within seconds
    mixing, dilute it (1:5) with sucrose/phosphate buffer and try again.
11  Shake the 'sediment' extract well and place a drop on a microscope
    slide under a coverslip. Examine under 'high power'. Repeat this wi
    the 'supernatant' extract and record the differences you observe.

*Questions*

a   *All the solutions contain phosphate salts which buffer at pH 6.5 an*
    *the sucrose/phosphate buffer contains $0.3 \text{ mol dm}^{-3}$ sucrose.*
    *Explain the importance of each of these ingredients.*

b   1   *Describe the changes in colour that took place in tubes 1 to 4.*
    2   *Explain how the controls help you to arrive at any conclusions.*

**c** *Compare the results in tubes 3 and 5, that is, the reactions with sediment and supernatant extracts. Which extract has produced more reducing activity?*

**d** *Explain why the enzymes which generate the reducing activity should be found mainly in one extract and not in the other. The results of your microscopic examination should help here.*

**e** *The rate of photosynthesis in intact green leaves can be limited by any one of the following factors:*
*1 Light.*
*2 Temperature.*
*3 Carbon dioxide.*
*One of these should have little effect on the production of reducing power in the leaf extract. Explain why.*

**f** *Describe how you might extend this investigation to show*
*1 how the production of reducing power depends on light intensity, and*
*2 how the production of reducing power varies with the wavelength of the light.*

## INVESTIGATION
## 7F The production of starch by leaves

(*Study guide* 7.4 'The reduction of carbon dioxide'.)

The presence of starch in leaves is often taken as evidence that photosynthesis has taken place. However, the presence or absence of starch may not always be a reliable indicator of photosynthesis. For example, potato tubers can form starch in total darkness and cannot be regarded as photosynthesizers. On the other hand, some plants, such as grasses, never produce much starch. Here we investigate some of the conditions that are required for the synthesis of starch in the leaves of a common plant. In particular, the investigation attempts to find out whether starch synthesis in these leaves is possible in total darkness.

*Procedure*
**1** Obtain healthy specimens of one of the following types of plant, enchanter's nightshade (*Circaea lutetiana*), tobacco (*Nicotiana* sp.), geranium (*Pelargonium* sp.), or busy Lizzie (*Impatiens walleriana*). Cut a few discs from the leaves with a cork borer.
**2** Test these leaves for starch as follows:
Immerse the discs in boiling water for a few seconds to break down cell walls and make them permeable.

Transfer to boiling ethanol (in a water bath) until all the pigments have been removed. *WARNING:* no naked flames.
Dip the discs in water for a few seconds to remove the ethanol. Immerse in a solution of iodine dissolved in potassium iodide. A blue or black colour indicates the presence of starch.
Provided the plants are healthy and have been well illuminated recently, starch should be present.

3   Put the plants in a light-proof box or cupboard for 48 hours. Again cut discs and test for starch. This time leave the decolorized discs in a solution of iodine dissolved in potassium iodide for at least ten minutes. If no starch is present proceed with the next stage. If starch is present, return the plant to darkness for twenty-four hours and test again.

4   Remove the gas from a solution of glucose (5 per cent in distilled water) by one of the following two methods:
(a) Half fill a filter flask with the solution. Connect the side arm of the flask to a good filter pump with vacuum tubing and place a tight fitting rubber bung in the top. Start the filter pump and keep swirling the solution in the flask for ten minutes, using a magnetic stirrer. *Before turning the pump off disconnect it from the filter flask.* Use the solution immediately and avoid shaking it while filling the syringes.
(b) Heat the glucose solution to boiling and then cool it down in a stoppered flask. Be sure that it has cooled to room temperature before using the solution. Avoid shaking it up while filling the syringes.

5   Cut discs from the leaves of the de-starched plant and use them to see if they are capable of producing starch in the dark as well as in the light. Float 5 to 10 discs, the right way up, on the surface of 5 per cent glucose solution in each of two syringe barrels open to the air (see *figure 27*).

6   If sufficient apparatus is available repeat step **5**, placing the leaf discs upside down on the surface of the solution.

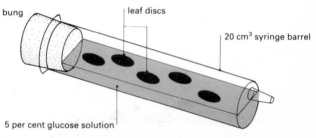

**Figure 27**
Floating the leaf discs in the syringe barrel.

**7** Remove the plungers from two more syringes and fit them with taps (see *figure 28*). Put the same number of leaf discs into each and add 15 cm³ of gas-free glucose solution. Tap the side of the syringes to dislodge any air bubbles from the discs. Replace the syringe plungers, open the taps and carefully push all the air out of the syringes. Close the taps.

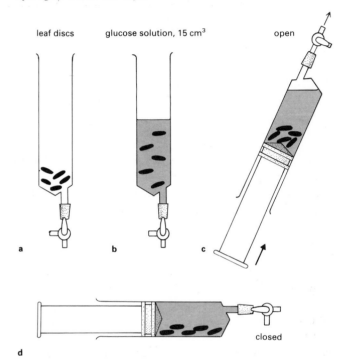

**Figure 28**
Placing the leaf discs in a gas-free environment.

**8** Put one vessel from step **5**, one from step **6**, and one from step **7** in the light. Place the remaining vessels in the dark. Leave for twenty-four hours. (Make sure that all the syringes are properly labelled.)

**9** Test the leaf discs for the presence of starch (see step **2**) and record the results in a suitable table. (Again, immerse the leaf discs in iodine solution for ten minutes.)

*Questions*

**a** *In which conditions did leaf discs produce starch?*

**b** *What are the leaf discs deprived of when they are immersed in the solution in step 7?*

c   *Under what conditions, if any, did your leaf discs synthesize starch in the dark? Propose a hypothesis to account for this result. How would you test your hypothesis?*

d   *Do the results of the investigation affect the validity of the starch te. as an indicator of photosynthetic activity? If so, give details.*

## INVESTIGATION
## 7G   Carbon fixation in CAM plants

(*Study guide* 7.6 'Crassulacean acid metabolism – the CAM pathway'

Photosynthesis is a complicated process involving several metabol pathways:

1 Light energy is trapped by the chloroplast and is converted into the chemical energy of ATP and NADPH.

2 Gaseous carbon dioxide is 'fixed'. This means that it is incorporated into an organic acid, such as phosphoglyceric acid or oxaloacetic acid. This pathway consumes some of the energy trappe previously.

3 Once 'fixed' the carbon atoms are next incorporated, at the expense of more energy, into the great variety of organic compound needed by the plant. This often involves the synthesis first of a suga such as glucose, and perhaps temporary storage as starch.

   In most plants all of these processes occur together. The overall rate photosynthesis will be governed by the slowest process. In bright sunshin this may be the fixation of carbon dioxide, which is limited by t availability of this gas. (Its concentration is only 0.03 per cent in t atmosphere.) The rate of gaseous exchange may therefore be a cruci factor controlling the rate of photosynthesis. In other words, to take fu advantage of bright sunshine a plant would have to keep its stomata wic open. However, this may sometimes prove to be a disadvantage becau water loss, by transpiration, may exceed water uptake through the root especially in arid conditions. There are a number of plants which use qui a different strategy to combine the different processes which make u photosynthesis. They are generally plants which are adapted to life in d places (xerophytes). They keep the 'fixation' of carbon dioxide qui separate from the energy-consuming reactions which utilize the 'fixe carbon.

   In 1947 a botanist, Meirion Thomas, claimed that certain plants the family Crassulaceae open their stomata only in the dark. H suggested that they fix carbon dioxide during the night by the followir metabolic pathway:

$$CO_2 + \text{phosphoenolpyruvic acid} \rightarrow \text{malic acid}$$

Without a source of energy further reaction would not be possible. Therefore the organic acids that are the initial products of carbon dioxide fixation, should accumulate in the leaves. According to this scheme, when light energy is available it will be used to convert the carbon, fixed in the dark, into carbohydrate. Photosynthesis could then proceed during the day with closed stomata and little or no gaseous exchange. The proposed sequence of reactions was called 'Crassulacean acid metabolism', CAM for short. Though first discovered in this family it occurs in plants from a wide variety of other families.

The aim of this investigation is to test the predictions:

**1** that acidic compounds accumulate in the leaves of certain plants while they are in the dark; and

**2** following this, that exposure to light leads to a reduction in the levels of acid in these plants.

*Procedure*

**1** The investigation described below should be carried out on two species of plant for comparison. One should be a CAM plant, for instance *Crassula argentea* (Jade plant). The other should be a non-CAM plant, for instance ivy or grass, as a control.

*Pretreatment of plants*

**2** Place the plants to be studied in a light proof container for 12–18 hours. If possible the atmosphere inside the container should be enriched with up to 1 per cent carbon dioxide.

**3** Expose half of the plants (the 'light' plants) to bright sunshine for 4–8 hours. The remaining plants (the 'dark' plants) are left in the dark until just before the extract is prepared.

**4** The class should have the following groups of plants prepared for study:

| CAM species | | non-CAM species | |
|---|---|---|---|
| light treated | dark treated | light treated | dark treated |

*Preparation of extract*

**5** Prepare separate extracts from the 'light' and 'dark' plants. (To save time each group of students can prepare an extract from one plant only. The different groups should compare results, but to make this possible each group must follow the same procedure in every detail.)

**6** Accurately weigh 10 g of leaves and chop them into small (1 cm$^2$) pieces with scissors.

7    Grind the leaves in a pestle and mortar with 30 cm$^3$ of distilled wat[er]
     and a little fine quartz sand.
8    Filter the homogenate (liquid) through fine gauze and divide it
     equally into two centrifuge tubes. Spin these in a bench centrifuge a[t]
     maximum speed for three minutes.
9    Carefully pour the clear extract into a test-tube.

*Measurement of the acidity of the extract*

10   Calibrate a pH meter by placing the electrode in solutions of know[n]
     pH. (One standard solution with a pH between 4 and 7 will do.
     Ideally two standard solutions should be used, one at about pH 4
     the other at about pH 7.) Follow these precautions:
     **1** Take great care to avoid knocking the glass electrode.
     **2** In between placing the electrode in different solutions, wash it by
     dipping it into a beaker of distilled water.
11   Measure the pH of the extracts from 'light' and 'dark' plants.
     *OR*
12   Fill a burette with the standard solution of sodium hydroxide.
13   Accurately measure 10 cm$^3$ of one extract into a small conical flask.
     Add three drops of phenolphthalein indicator.
14   Note the initial burette reading. Run the solution of sodium
     hydroxide into the conical flask drop by drop. Swirl the liquid.
15   When the phenolphthalein turns pink note the volume of sodium
     hydroxide solution used. Take care this end point is not 'overshot'.
16   Repeat the titration with a second and, if possible, a third 10 cm$^3$
     sample of the same extract.
17   Titrate duplicate or triplicate samples of the other extracts.

*Questions*

a    *Which extract ( from 'light' or 'dark' treated plants) contains a
     higher concentration of acids?*

b    *Is it possible to decide whether the measured difference in acidity
     between the two extracts is significant?*

c    *Is the result of this investigation consistent with the scheme for
     Crassulacean acid metabolism that was outlined in the introductio[n]
     Explain your answer.*

d    *What advantage might this type of metabolism give to a plant
     growing in a dry habitat?*

e    *When water is readily available, CAM plants can be at a
     disadvantage in competition with non-CAM plants. Give reasons fo[r]
     this disadvantage.*